PROBABILITY, RANDOM VARIABLES, AND RANDOM SIGNAL PRINCIPLES

McGraw-Hill Series in Electrical and Computer Engineering

Senior Consulting Editor

Stephen W. Director, *Carnegie Mellon University*

Circuits and Systems
Communications and Signal Processing
Computer Engineering
Control Theory
Electromagnetics
Electronics and VLSI Circuits
Introductory
Power and Energy
Radar and Antennas

Previous Consulting Editors